25th Topical Conference on Refinery Processing 2022

Topical Conference at the 2022 AIChE Spring Meeting and
18th Global Congress on Process Safety

San Antonio, Texas, USA
10 – 14 April 2022

ISBN: 978-1-7138-6819-4

Printed from e-media with permission by:

Curran Associates, Inc.
57 Morehouse Lane
Red Hook, NY 12571

Some format issues inherent in the e-media version may also appear in this print version.

AIChE
120 Wall Street, FL 23
New York, NY 10005-4020

Phone: (800) 242-4363
Fax: (203) 775-5177

www.aiche.org

Additional copies of this publication are available from:

Curran Associates, Inc.
57 Morehouse Lane
Red Hook, NY 12571 USA
Phone: 845-758-0400
Fax: 845-758-2633
Email: curran@proceedings.com
Web: www.proceedings.com

TABLE OF CONTENTS

639641 Digitalization's Role for Optimizing Sustainability

Abstract:

Operating our current assets in the most sustainable way possible is critical. Societal concerns, government regulations, investor activism, and customer demands must be addressed with real action. While bold statements and aggressive targets may be sufficient in the short term, specific action steps to drive progress is necessary.

Achieving zero emissions or a net zero carbon foot print is not easy. The technology and infrastructure to do this profitably does not exist currently. Any profitability expectations would be based on speculative assumptions of advances in technology and infrastructure as a result of universally applied government regulation across industry sectors that change the competitive landscape and profitability profile. Designing, constructing, and starting a refinery or petrochemical plant to operate in this manner would take 6 to 8 years and then need expectations of running for 20 or 30 years as a minimum. So a high level of investment would be required now to achieve the required reduction a decade from now. Certainly available options (i.e. feedstock changes, green/blue hydrogen or green electricity for heat, CCUS, and others) should be considered and prices understood so that investment can be made when confidence in profitability can be established.

In the interim however, we can commit to operating our existing assets in the most efficient manner possible. As chemical engineers we are trained to optimize production by addressing efficiency, throughput, yield and availability. Digitalization provides the opportunity to utilize this knowledge and domain expertise in a manner that presents a view of operations where baselines are established, tradeoffs between energy efficiency and production are understood, and emissions can be minimized. Measurement and awareness are the first steps to achieving operational excellence in these areas.

Digitalization provides measurement, understanding and insight so that actions can be taken to improve sustainability. Complete heat and material balances around the facility can be included in a visual manner to show operational performance for emissions, energy efficiency, and production. Live steam balance can be presented to show utility tradeoffs possible. The real time operational tradeoffs can be presented to operations for tuning per the plant's needs.

This can be extended to the strategic decisions at an engineering and management level where operational strategy for running a plant change as prices change. Pinch analysis opportunity identification for the heat exchanger network will vary as the tradeoff between capital expenditure and energy costs vary. Thresholds and investment profitability will change as cost profiles change. The profitability of changing feedstocks and substituting hydrogen for fuel gas will change as market conditions and regulation is implemented. Being prepared and understanding the opportunity can be included as part of the digitalization solution.

Case examples of digitalization being applied to enhance decision making on a real time basis to understand current operations, opportunity for improvement, and recommended steps needed to be taken will be reviewed.

639412 Mixed Reality for Predictive Maintenance of Heat Exchangers

Abstract:

As a part of the digital transformation of the process industry, Mobile Operator Rounds are carried out daily as a way of monitoring a facility's operational performance.

In this interactive presentation, we will discuss the benefits of using mixed reality as a part of Mobile Operator Rounds for monitoring the performance of heat exchangers. The benefits of this technology for refinery applications will be highlighted.

639653 Live Column Hydraulics - How Does It Help?

Abstract:

Overall Manufacturing Asset Effectiveness is significantly affected by availability of an equipment and it's ability to perform to level of operational expectations. In refinery process, the capability of the overall plant effectiveness is significantly governed by the effective utilization of individual column capacities while ensuring reliable operation i.e. maximize capacity while ensuring on-spec product with optimal product recovery. In addition to this, variable feedstock quality and product pricing adds to the complexity as more than often this governs the overall plant capacity and profitability.

Traditional approaches more than often lead to setting up of conservative limits to ensure reliable operation. With the advancement of big data and solutions focused on AI/ML techniques, the power of analytics can be effectively used to improve column capacity while ensuring reliable operation i.e. ensure on-spec production, maximize product recovery and avoid unplanned downtime à "finding problems before they find you". Using the power of data analytics ensures that the advantages from process optimization and advanced controls are consistently realized.

More than often the symptomatic indicators reside in the large quantity of a variety of data that are measured on a continuous basis and archived in the plant historian. It would need a rigorous, and time-consuming exercise to review these wide variety of millions of data points - collected on a minute wise or even higher frequencies – to identify such symptoms and this is difficult for an unaided human mind, unless it shows up obviously on a commonly monitored parameter. Several of these parameters measured are non-linearly related to one another, further complicated by different types of errors in the collected data.

Big Data Analytical techniques can be utilized to effectively analyse hundreds of minute wise tag data for a process equipment, in real time. This paper describes how Machine Learning Techniques in combination with first principle models have been applied to Distillation Column operation in an Manufacturing unit to provide useful, non-intuitive insights that are usually hidden in the large volumes of data collected. This paper covers the case studies covering applied machine learning techniques for troubleshooting and improving Distillation Column performance. How Digital Twin model is deployed with real-time analytics options providing augmented intelligence to maintain operations at best possible

Case Study 1: Column – Live Hydraulics: A real time analytics solution was developed for providing live hydraulics, tray loadings at different sections of C2 Splitter Column.

- Intuitive insights to address capacity constraints by addressing flooding areas with smart changes in operating parameters
- Ensuring reliable operation, tracking PDI and other critical instrumentation

Case Study 2: Column – Real time Optimizer: A multivariate ML model based on live data provided real time guidance to plant operations to ensure on-spec production without limiting column capacity or overall performance.

These and other case studies would be presented to demonstrate how power of analytics can be used to offer real time decision insight to operations, driving actionable outcomes creating new levels of operational excellence.

641678 Sponsored Technology Workshop - Enhance Reliability of Refinery Operations with Electrolyte Based Process Simulation Solutions - OLI Systems

Abstract:

Electrolyte thermodynamics play a crucial role in the material reliability of a refinery asset. As refineries focus on making products on targets, maximizing profits and reducing feedstock costs, asset reliability may be negatively impacted as we push operational boundaries. Attend this workshop to understand how chemistry insights can help increase reliability and extend asset life by setting Integrity Operating Windows.

During this workshop, the presenter will take a focused look at the chemistry of contaminants in crude unit overhead systems, and will, in detail, explain the phenomena that causes chemical corrosion and damage in the units. We will investigate the reasoning behind neutralizing amine and wash water injection points and their separation.

This workshop will deliver actionable insights on why chemical companies suggest particular neutralizing amines for the overhead systems and how refiners can use the same tools to assess the performance of chemical additives. Finally, we will review a real-life case where an opportunity crude was evaluated as an excellent source to increase profitability. However, the elevated level of contaminants in the crude caused the corrosion rate on the overhead, as measured by boot water analysis and iron content, to increase dramatically.

In this workshop we will cover:

- Phase change corrosion: Ionic Dew Point (sometimes called Initial Dew Point), Water Dew Point and -Hydrocarbon Dew Point Calculations
- Neutralizing amine selection criteria
- Wash water rate optimization
- Processing of high-contaminant crudes and desalter upsets
- Developing Integrity Operating Windows

640661 How to enable high yields from bio-feedstock with the ExxonMobil Renewable Diesel (EMRD^{TM}) Process Technology

Abstract:

The ExxonMobil Renewable Diesel (EMRD^{TM}) process technology was recently launched to provide a two-stage hydroprocessing unit solution for production of renewable fuels from bio-based feedstocks. These feedstocks can include vegetable oils, used cooking oils, or animal fats. The first stage of EMRD converts the triglyceride feedstock to n-paraffins through predominantly hydrodeoxygenation chemistry. The second stage contains ExxonMobil's Bio-Isomerization Dewaxing (BIDW^{TM}) catalyst to selectively isomerize the n-paraffins to iso-paraffins, thereby improving cold flow properties. The selectivity of the dewaxing stage is crucial in maintaining yield. Hydroprocessed bio feedstocks contain a higher percentage of n-paraffins than conventional hydroprocessed distillate, and therefore require a higher severity dewaxing operation to target comparable cloud point properties. The high severity operation often introduces cracking chemistry that can reduce yield; high selectivity allows cold flow improvement while maintaining molecules in the jet and diesel boiling range. The EMRD process harnesses the yield benefit provided by BIDW with an efficient process design based on ExxonMobil's long history as an owner and operator of hydroprocessing technologies.

639977 The Future of Delayed Coking, Expanding Flexibility Will be the Key.

Abstract:

Thermal conversion and batch processing may not be as exciting as high pressure, ebbulated bed catalytic units or organometalic alkylation but it makes up for this in flexibility and potential uses for the 21st century. The presentation will review the history of delayed coking as well as the future uses of the workhorse unit in refining. There are no other process units in refining that can operate at 35-120+% of rated capacity or can take feedstocks that vary from basically whole crude to bottomless light sweet to waste streams from other processes. In the future, the potential exists to support the battery market and chemicals intermediates with this type of unit as well. The future of refining will include a significant amount of rationalization of capacity and sites. If you want your refinery to survive and thrive into the foreseeable future, a delayed Coker is the price of admission. The presentation will explore the range of operations possible now and into the future.

639015 HF Alkylation: Novel Thermodynamic Model to Simulate Phase Change Corrosion Zone

Abstract:

HF Alkylation: Novel Thermodynamic Model to Simulate Phase Change Corrosion Zone

Ezequiel Vicent, Peiming Wang, Andrzej Anderko

OLI Systems, Inc. 2 Gatehall Drive, Suite 1D, Parsippany, NJ 07054

email: Ezequiel.Vicent@OLISystems.com

INTRODUCTION

The HF alkylation unit performs a series of specific chemical and physical processes, catalyzed by hydrofluoric acid (HF), to convert isobutane and light olefins (i.e., propylene and butylene) into alkylate, a mixture of larger branched-chain alkanes that are mostly isooctane and isoheptane. Alkylate from this process has a high Research Octane Number (RON of 94), a low Reid Vapor Pressure (RVP of 4 psi), and very little to no sulfur. This allows a typical refinery to improve their gasoline blends to reach premium and mid-range grades, while lowering the overall products' sulfur content and RPV point.

HF alkylation units, however, are notorious for having significant carbon steel corrosion issues, particularly in the feed lines from the acid settler to the isostripper (or depropanizer, depending on unit configuration), as well as overhead lines. This corrosion has been caused by the formation of a water-enriched acid phase (rich-HF or RHF) that leaves the settler as entrained acid in the liquid hydrocarbon. Specifically, the most corrosive environments have been created by the formation of a water-rich RHF phase when condensing out of a vapor or when reaching solubility limit in liquid hydrocarbon stream. Corrosion from this water-rich HF can damage the piping systems and equipment, causing dangerous accidents. Thus, a thermodynamic model that can provide accurate prediction of phase equilibria in the HF alkylation environment is essential for predicting and mitigating corrosion.

DESCRIPTION OF THE WORK

An industry sponsored Joint Industry Project (JIP) was formed to enhance our understanding of the relationship between the rich-acid and corrosion through the development of a novel electrolyte thermodynamic model.

Thermodynamic modeling provides a foundation of process simulation to predict phase equilibria and their changes in the mixtures involved in HF alkylation so that optimal operating conditions can be defined to prevent water-rich HF phase changes in locations where corrosion is prone to occur (e.g., outside heat exchangers). In this work, OLI's mixed solvent electrolyte (MSE) thermodynamic model

[1] has been applied for this purpose. The MSE model defines properties of each component and interactions between any two components in the HF alkylation mixtures including hydrocarbons, HF, water, selected fluorocarbons and representative acid soluble oils (ASOs). Appropriate definition of these interactions allows accurate prediction of phase equilibria under varying conditions of temperature, pressure, water content in HF, etc. Consistent thermochemical properties of all species / components make it possible to determine chemical speciation that provide insights into corrosion phenomena in the alkylation process.

RESULTS AND DISCUSSION

The key to assess corrosion risks in the HF alkylation process is to predict how phase equilibria (liquid-liquid and vapor-liquid equilibria) vary as conditions change. Thus, the focus of the model development is to provide accurate prediction of solubilities in the system H2O-HF-hydrocarbons-acid soluble oils. Selected model results and their comparison with literature data will be shown during the presentation for the solubilities of HF in hydrocarbons. Excellent agreement has been obtained between the calculated solubilities and experimental data. These results show how the solubility of HF vary with temperature for a representative list of components found in the settler hydrocarbon fluid. These results can be useful for the effective modelling of HF solubilities in the hydrocarbon phase and to understand the phase equilibria of the two-phase fluid (an electrolytic acid phase and a hydrocarbon phase) that leaves the settler.

More importantly, the solubility effects of HF vs H2O in hydrocarbon drive the corrosivity of the electrolytic solution. The MSE thermodynamic model can predict the solubility differences under variable conditions and the effect of water on HF solubility in hydrocarbons can also be predicted. Model results show of solubilities of HF and water in the a variety of hydrocarbons (comprising the hydrocarbon effluent of an acid settler), showing that HF is more soluble than water in hydrocarbons. Model results suggest that when the liquid hydrocarbon is saturated with HF upon heating, the water-to-HF ratio in the RHF phase will increase because more HF than water will dissolve in the hydrocarbon phase. This results in an increased relative water concentration in the RHF phase, leading to an increasingly corrosive environment. Through the JIP, appropriate interaction parameters between mixture components were developed that allow the MSE model toprovide quantitative prediction on the variation of water content in the entrained RHF phase as temperature increases at any given stream conditions (i.e., composition and pressure).

The model was applied to a JIP member acid settler effluent, simulating the heating process encountered in the feed/rcy-iC4 heat exchangers (preheaters) prior to fractionation, on the main feed line. The model shows that as we increase the fluids temperature, HF will dissolve into the hydrocarbon phase with greater affinity than H2O, increasing the H2O relative concentration in the rich-acid phase. The increased water concentration in the acid causes an increase in the hydronium ion of the rich-acid phase, thus, increasing its corrosivity. The water relative concentration in the RHF will continue to increase (increasing the electrolytic phase corrosivity) until a temperature is reached where all the RHF dissolves into the hydrocarbon phase. We call this temperature the Transition Temperature (TT). The model and its calculation of the TT can help refiners develop Integrity Operating Windows (IOW) to ensure the TT is reached at the heat exchangers to avoid corrosion in the main feed line to the main fractionator.

The new thermodynamic model developed in this work can be combined with process simulation to provide much improved understanding of corrosion caused by RHF phase change in HF alkylation fractionation systems. In addition to the aforementioned application for the feed line to the main fractionator, this model can be applied to the depropanizer, HF stripper, tower flash zones and overhead lines where phase changes are prone to occur. Refiners can use this model to calculate a fluid's TT and ensure that operations happen above this temperature to avoid corrosion, and create IOWs for the fractionation system of the unit.

This presentation will show the model's thermodynamic foundation and results obtained in predicting phase equilibria in the HF alkylation environment. An example will be demonstrated on how refiners can make use of this model to develop integrity operating windows to avoid phase-change corrosion in the HF alkylation unit.

REFERENCES

1. Wang, P.; Anderko, A.; Young, R. D. 2002, Fluid Phase Equilibria, 203, 141-176
2. Chen, X.; Lee, C. ; Wu, Z. 1994, AIChE DIPPR Data Ser. 28.
3. Simons, J. H. 1931, *Amer. Chem. Soc.,* 53, 83-87.
4. Marcus, Y.; Shamir, J.; Soriano, J. 1970, *Phys. Chem.,* 74, 133-139.
5. Butler, E. B.; Miles, C. B.; Kuhn, C. S. 1946, *Eng. Chem.,* 38, 147-155.
6. Kang, Y. W. 1999, *Fluid Phase Equil.* 154, 139-151.

640167 Fundamentals of Hydrocracking Operation in Oil Refinery

Abstract:

The goal of this paper is to present basic fundamentals of hydrocracking operation in a refinery. Different types of feed stocks and the purpose of hydrocracking is explained with sample feed stocks. The role of hydrocracker in the grand scheme of refinery process and petrochemical integration is explained using multiple sketches. Details of hydrocracking reactions involved in the reactors is explained. Different reactor stages i.e., single stage vs two stage options are discussed with advantages and disadvantages. Different operating conditions and common operations problems are explained with examples. Potential limitation of the process is discussed. The role and importance of fractionator in the hydrocracker unit complex is discussed. The hydrogen requirement, the recycle gas compressor, and oil-to-gas ratio is explained in relation to the product yield. Safety aspects of the process is touched briefly.

639588 Crude Unit Heat Integrationa Case Study

Abstract:

Crude Unit Heat Integration

A Case Study

Andrew W. Sloley

At a minimum, crude oil will remain the source of transportation energy through 2050, and perhaps beyond. Crude oil distillation is the first step in converting crude into streams for further processing into transportation fuels.

Crude oil distillation is energy intensive – all the desired products are distillates that must be vaporized. Modern units use complex heat integration to recover the energy from condensing the product and reflux and using it to heat the feed. Extensive academic and industry work has gone into optimizing the heat integration for maximum energy efficiency and minimum investment. However, finding the optimum configurations remains complex and time consuming.

This is even more true for revamping units. Revamps add additional constraints to the energy analysis. This may make what appear to be obvious energy integration steps impractical. A case study for a recent revamp is presented that balances multiple factors to find a profitable and more efficient heat integration configuration.

639480 Heat Exchanger Network (HEN) Retrofit for a Sustainable Energy Efficient Design in Aramco's Lube Refinery

Abstract:

Industrial sector is accountable for approximately one-third of world-wide energy usage that relies on fossil fuels as the primary source of energy. A significant percentage of energy consumed is spent to supply utilities for oil & gas facilities so that energy could be engendered for other sectors. The adverse impact of fossil-fuel combustion has instigated efforts to minimize carbon emissions, whilst environmental regulations have a substantial effect on the energy's production cost. Furthermore, it is a competitive world with ever-increasing energy prices, industrial communities are convened to brainstorm over conservation of energy resources and profitability. Energy efficiency should be a foundational strategy to support plans to meet the climate policy goals as well as for being lucrative. Energy efficiency with its integral role in energy intensive refining processes is critically essential to achieve desired energy-saving potential and reduction in CO_2 emissions. Pinch technology imparts a systematic technique for saving energy in processes and on entire sites. This technique helps optimize thermal heat recovery, and immediately became popular as a theoretically elegant yet feasible approach to design Heat Exchanger Networks (HENs). The paper illustrates, through a case study, the 'plant lifetime retrofitability' concept in a standard oil refinery's VDU plant. The study demonstrates the impact of energy demand reduction by identifying retrofit projects for a Vacuum distillation unit (VDU), utilizing retrofit with retrofit in mind approach. The case study's objective is to look for the most inexpensive retrofit projects via arranging each initiative in phases for implementation using the existing topology and available capital investments. Each energy saving' initiative is implemented in a way that does not hinder the future identified initiatives. The analysis was conducted with a commercially available Heat-Int software, with an initial minimum temperature difference (ΔT_{min}) of 34°C with corresponding hot and cold utility requirements of 21.53 Gcal/hr and 17.13 Gcal/hr, respectively. The analytical outcomes disclosed a 2.02 Gcal/hr energy reduction in the vacuum distillation unit (VDU) with simultaneous generation of MP and LP steams of 4.4 and 3.8 tonnes/hr, respectively. The proposed modification can be implemented in three phases, execution of each phase retrofit won't hinder the future retrofits and beyond.

639506 Heavy Crude Oil Processing to Improve Pipe Line Transportation Properties

Abstract:

Rationale

Transporting heavy crude oils through pipeline or by other transportation means is challenging and often requires the petroleum industry to resort to rail transport for inland sources, adding large volumes of diluent, or adding very high energy input, not to mention the extra handling and danger of spillage.

Objective

Describe a novel approach to economically transporting heavy crude oil by pipeline over long distances by reducing the viscosity of heavy crude oils using low-concentration chemical addition and integral mechanical treatment. The treated heavy crude oils do not require heating over 40°C, adding large volumes of diluent, or adding large volumes of chemicals to promote transportation through pipeline.

Methods

'EZ Flow' combines hydrodynamic cavitation processing of the heavy crude oil with addition of proprietary, low concentration chemical formulations (< 2% by volume) to promote the reduction of the heavy crude oil's viscosity. First, over thirty different chemical formulations were rapidly screened at the bench-scale with sonication and mixing as the mechanical mixing method. Second, the most promising chemical formulations identified at the bench-scale were scaled up to a small, 15-gallon pilot plant scale hydrodynamic cavitation system to process heavy crude oil and chemical formulation and to collect samples during and after processing for analyses.

Results

Chemical formulations 30, 31, and 32, treated using the 15-gallon hydrodynamic cavitation pilot scale system, yielded viscosity reduction of Belridge heavy crude oil (API° 11) by up to 60%.

Conclusion

'EZ Flow' reduced a Belridge heavy crude oil's viscosity up to 60%, making this technology a candidate for converting this type of heavy crude oil into a flowable material suitable for transport through existing pipelines at moderate temperatures. If commercialization continues as planned, the technology could result in significant decreases in transportation costs and potentially limit environmental impact for oil producers. There is also a high probability that this technology could be used for transporting other unconventional oils. Also, the treated heavy crude oil could be stored for extended periods of time (months) without losing its reduction in viscosity. EZ Flow technology is filed with the U.S. Patent Office in 2021, U.S. Patent Application Serial No.: 17/452,458.

638807 Radio Frequency Driven Heating of Catalytic Reactors for Portable Green Chemistry

Abstract:

Here, we propose a novel integration of radio frequency (1-300 MHz) responsive nanomaterials with conventional catalytic materials to realize a new class of heterogeneous catalysts that undergo uniform volumetric and localized heating to drive chemical transformations at the modular scale. Approximately 80% of chemical manufacturing involves heterogeneous catalytic reactions, which currently require heating via steam utilities or fired furnaces, and thus contribute to global greenhouse gas emissions while also limiting distributed chemicals production. Our approach uses an electric route to produce chemicals where radio frequency (RF) electric fields and their interaction with carbon/ceramic nanomaterials are utilized to selectively heat the catalyst composition. A proof-of-concept is demonstrated using the commonly studied methanol steam reforming reaction on a platinum catalyst. In this study, two RF susceptors were used: carbon nanotubes and silicon carbide fibers. The conversion rate of methanol using RF heating was comparable to oven heating at varying temperature and catalyst combinations. This is a potential breakthrough over conventional catalytic reactors in that it enables small, safe, sustainable, on-site, and on-demand production of chemicals in the absence of traditional manufacturing infrastructure.

639330 Hydrotreating of Bio-Oil Model Compounds: Experimental Results and Kinetic Modeling

Abstract:

Introduction

Although bio-oil from fast pyrolysis of lignocellulosic biomass is a potential source for fuels and valuable chemical building blocks, such bio-oils cannot be directly integrated into existing petroleum refining facilities. Indeed, fast pyrolysis bio-oil has an extremely complex organic character with not only a high amount of carbon, but also large quantities of oxygenated species. Since the presence of oxygen reduces the quality of bio-oils as a fuel source, bio-oils require a dedicated pre-refining step, which is generally a hydrotreating process. However, due to its high complexity, in both molecular and reaction terms, the upgrading of bio-oil reveals to be difficult. To correctly design such a process, one must acquire insight in the molecular composition of bio-oils and of the reactivity of the various species.

The present work describes a methodology for modeling the upgrading of bio-oils. The proposed method will be applied to construct kinetic models for the hydrotreating of bio-oil model compounds. This strategy will enable a further understanding of bio-oil reactivity, opening the doors for the modeling techniques for full-range bio-oils.

Methodology

As modeling strategy, a direct discrete simulation of reaction events was chosen. The approach is based on a kinetic Monte Carlo (kMC) method, termed Stochastic Simulation Algorithm (SSA)[1,2], and describes the evolution of a system, molecule by molecule. Unlike the classic procedures, this method can describe the evolution of a molecular mixture over time without a pre-defined reaction network.

The stochastic approach is based on the identification of all possible reaction events at a given moment in time and on the calculation of the reactivities for each event. Thanks to the work of Gillespie[1,2] and recalculating apparent first order rate coefficients for the reactions, the reaction probability can be generated for each reaction event. By assembling all the probabilities of all molecules, the overall reactivity allows to determine, via Monte Carlo sampling, the time interval until the next reaction, and to select a given reaction. The sampling process of both variables is then repeated until the final simulation time is reached.

The SSA requires as inputs a list of reactant molecules, the different types of reactions that can occur and the kinetic parameters associated to these transformations. Due to its stochastic nature, this method generates a different outcome for each simulation, which does not allow to construct the general evolution of the reaction system over time. Therefore, as with any stochastic approach, the method has to be based on several simulations, each of which calculates discrete time trajectories of molecular populations. Their average profile will converge to the same results as deterministic methods.

Results

To understand the reactivity, the hydrotreating of two model molecules was studied: Guaiacol and Furfural. The experimental data was obtained by Ozagac[3] and by Costa da Cruz[4] for the hydrotreating of guaiacol and furfural over a NiMo/Al_2O_3 catalyst in a batch reactor at 13 MPa, at 200°C, 250°C, and 300°C, and for reaction times of 3h and of 5h. For several experimental conditions, blank runs were performed and showed a relatively high conversion. For guaiacol, an initial conversion of 10% was obtained at 300°C after the heating up and cooling down the batch reactor.

For the kinetic modeling of hydrotreating guaiacol and furfural, the reactions for hydrogenation, dehydrogenation, demethylation, demethoxylation, decarbonylation, decarboxylation, deoxygenation, transalkylation, and dehydration were implemented in the Stochastic Simulation Algoithm. For the model molecules, the SSA method is able to predict the expected trends, not only for the conversion of the reactants, but also for the generation of the various products and for the hydrogen consumption. Additionally, this approach also predicts the transformation of the several reaction products into smaller molecules. Due to the high reactivity of guaiacol, the heating and cooling periods contribute significantly to the overall conversion. Hence, the reactor simulations need to account for these dynamic periods.

The model results were compared to the experimental data. Accounting for the thermal behavior of the reactor during heating, reaction and cooling, the kinetic parameters of the reactions were estimated. A good agreement is observed. The comparison of the simulated and the experimental selectivities for methanol and dialcohols shows that the model predicts, within the error of the experimental data, the correct order of magnitude for both chemical families.

Conclusions

The SSA methodology was applied to bio-oil model molecules and showed the correct trends when compared to available experimental data. Through this study, a kinetic model was generated that is able to simulate the reactions of guaiacol and furfural under hydrotreating conditions. It was shown that the SSA method is an excellent simulation tool for complex reaction networks without the need for a pre-defined reaction network. Furthermore, this method enables a better understanding of these systems, making it suitable to be applied to a full range bio-oil.

Literature Cited

1. Gillespie D.T., "A general method for numerically simulating the stochastic time evolution of coupled chemical reactions," *Journal of Computational Physics*, **22**(4), pp. 403–434 (1976).
2. Gillespie D.T., "A rigorous derivation of the chemical master equation". *Physica A: Statistical Mechanics and its Applications* **188**(1-3), pp.404–425 (1992).
3. Ozagac M., "Etude mécanistique de l'hydroconversion catalytique de bio-huiles de pyrolyse," PhD thesis, Université Claude Bernard - Lyon 1 (2016).
4. Costa da Cruz A.R., "Compositional and kinetic modeling of bio-oil from fast pyrolysis from lignocellulosic biomass", PhD thesis, Université Claude Bernard - Lyon 1 (2019).

639344 Computational Characterization of Pore Network Models for Gamma-Alumina Catalysts

Abstract:

Introduction

Understanding how transport phenomena and chemical reactions within porous catalysts are affected by the texture of the support opens the door to optimizing the catalyst performance. New and more rigorous heterogeneous models for chemical reactor design could be created by accounting explicitly for the phenomena occurring at the pore scale. To model structures of porous catalysts several approaches can be found in literature [1,2]. Generating computational structures that match structural descriptors of the real material requires information about the textural properties and the topology of the material. In this work, the impact of the porous structure of gamma-alumina catalyst supports on mass transfer phenomena in non-reactive and reactive systems is studied. The representation of the alumina support start from previous work [3], in which a network of interconnected cylindrical pores is created by means of a stochastic assembly algorithm. By using a hierarchical network, Ferreira [4] was able to computationally generate several pore network structures whose textural and transport properties matched those obtained experimentally in Kolitcheff's work [5]. The objective is now to further constrain the number of possible structures by generating a network that allows representing experimental results obtained from different characterization techniques. Nitrogen Porosimetry (NP), Mercury Porosimetry (MP), Nuclear Magnetic Resonance Cryo-Porometry (CP) and Thermo-Porometry (TP) are such characterization techniques that deliver quantitative and qualitative information about the porous structure by using theoretical and empirical models: pore size distribution, pore volume, and surface area are textural properties estimated by these techniques [6].

Methods

The pore networks used for the simulations were generated through a discrete reconstruction model based on a stochastic algorithm described by Ferreira [3,4]. For the models, two cases are considered: in the first case, the pores are modelled as cylinders that are connected through zero volume nodes, while in the second case the pores are cylindrical and connected to spherical nodes. The cylinders are organized in space using a 3D cubic lattice that can be regular or disordered. By using a Monte-Carlo approach, the creation process of every pore is totally random, constrained just by the pore existence probability defined as input.

To characterize the digital pore network, simulation algorithms for NP, MP, CP and TP were developed. The algorithms account for percolation phenomena and confinement effects occurring within the porous network during the characterization. Selection of the right computational algorithms to model the physical phenomena and to store and handle the network's information is quite relevant to limit the execution time. An Invasion Percolation (IP) algorithm was adapted for each modelled technique based on the work of Wilkinson and Willemsen [7] and the improvement made by Masson and Pride [8].

Finally, the pore network model representation of a gamma alumina has been used for simulating intraparticle diffusion in the case of selective hydrogenation of butadiene.

Results and discussion

The textural properties of the digital pore network were calculated both geometrically and by processing the results of the simulated characterization techniques. Compared to the direct calculation of the textural properties by simply counting the numbers of pores, the direct simulation of the entire characterization profiles provides a better interpretation of the experimental results by analyzing the validity of the assumptions made by theoretical and/or empirical models used for the calculation of the material's textural properties from the experimental data. The pore network generation tool is based on a ***O(log(n))*** algorithm that allows to create a network with pores in only 46 seconds on a Dell desktop computer with a 3.5 GHz Intel Xeon E5 CPU and 16GB of RAM. For the characterization techniques, the use of a dynamic priority list to handle the exploration of the network is valid for all techniques, as the phase change and/or percolation phenomena occurring in every technique are constrained by the topology of the network. The algorithms are capable of reproducing typical hysteresis behavior observed in the experimental results for alumina catalyst supports. By versioning the IP algorithm to our needs and handling the priority list using diverse versions of binary heaps, we were able to maintain a good time performance for the different algorithms created for each characterization. The execution time for the characterization techniques corresponds to an ***O(nlog(n))*** behavior and takes about 25 seconds for pores. To match the textural properties of the digital pore network with that of an actual alumina support, the input variables of the network model are modified until the differences between the output textural properties and actual values are minimized. A good agreement was observed between the properties of the digital twins and the actual aluminas.

Using the pore network model representation, the intraparticle diffusion was simulated in the case of selective hydrogenation of butadiene. By varying the pore network properties, their effect on the effectiveness factor was investigated.

Conclusions

In this work, a pore network generation algorithm was used to represent gamma-alumina supports. To obtain the textural properties of such a digital pore network, an Invasion Percolation algorithm was adapted in order to characterize this structure by means of the computer-equivalent of nitrogen porosimetry, mercury porosimetry, NMR cryo-porometry, and thermo-porometry. Using the developed algorithms, it was possible to generate digital pore networks whose properties are in good agreement with the properties experimentally measured by different characterization techniques on actual gamma-alumina supports. Finally, the pore network model representation of a gamma alumina has been used for simulating intraparticle diffusion in the case of selective hydrogenation of butadiene.

References

1. Ye G., Sun Y., Zhou X., Zhu K., Zhou J., Coppens M.O., Chemical Engineering Journal, 329 (2017) 56-65.
2. Euzen P., Raybaud P., Krokidis X., Toulhoat H., Le Loarer J.-L., Jolivet J.-P., Froidefond C., In Handbook of Porous Solids, Vol. 3. Weinheim: Wiley-VCH Verlag GmbH, 2002, 1591-1677.
3. Carreira Ferreira S., 2018, Modelling of the intra-granular mass transfer within catalytic reactors. Université de Lyon (in English).
4. Ferreira S., Verstraete J., Jolimaitre E., Leinekugel-le-Cocq D., Jallut C., Computer-Aided Chemical Engineering, 40, Part A (2017) 91-96
5. Kolitcheff S., 2017, Approche multitechnique des phénomènes de diffusion en hydrotraitement de distillats. Université de Lyon.
6. Coasne B., New Journal of Chemistry, 40 (2016) 4078-4094.
7. Wilkinson D., Willemsen J., J. Phys. A: Math, 16 (1983) 3365-3376.
8. Masson Y., Pride S.R., Transport in Porous Media, 102 (2014) 301-312.

638444 The Active Site and Mechanism for Catalytic Hydrogenation on Metal Cation Catalysts Supported on Nu-1000: Insights from Experiments and Microkinetic Modeling

Abstract:

The mechanism of ethene hydrogenation to ethane on six dicationic 3d transition metal catalysts is investigated. Specifically, a combination of density functional theory (DFT), microkinetic modeling, and high
throughput reactor experiments is used to interrogate the active sites and mechanisms for Mn@NU-1000,
Fe@NU-1000, Co@NU-1000, Ni@NU-1000, Cu@NU-1000, and Zn@NU-1000 catalysts, where NU-1000 is
a metal–organic framework (MOF) capable of supporting metal cation catalysts. The combination of experiments and simulations suggests that the reaction mechanism is influenced by the electron configuration and spin state of the metal cations as well as the amount of hydrogen that is adsorbed. Specifically, Ni@NU-1000, Cu@NU-1000, and Zn@NU-1000, which have more electrons in their d shells
and operate in lower spin states, utilize a metal hydride active site and follow a mechanism where the metal cation binds with one or more species at all steps, whereas Mn@NU-1000, Fe@NU-1000, and Co@NU-1000, which have fewer electrons in their d shells and operate in higher spin states, utilize a bare
metal cation active site and follow a mechanism where the number of species that bind to the metal cation
is minimized. Instead of binding with the metal cation, catalytic species bind with oxo ligands from the NU-
1000 support, as this enables more facile H2 adsorption. The results reveal opportunities for tuning activity
and selectivity for hydrogenation on metal cation catalysts by tuning the properties that influence hydrogen
content and spin, including the metal cations themselves, the ligands, the binding environments and supports, and/or the gas phase partial pressures.

639610 Custom Clean Hydrogen Carbon Capture

Abstract:

In refinery applications, the vast majority of hydrogen is produced via steam-methane reforming. Typically, this hydrogen production emits a high degree of CO_2 into the atmosphere; however, through effective deployment of carbon capture technology, the process CO_2 can be captured for sequestration or utilization, turning this Grey H_2 into Blue H_2. With existing and upcoming governmental policy, Blue H_2 can be viable in the current economic environment. With careful selection of carbon capture technology and configuration, specific economic objectives can be optimized, further improving economic viability.

640304 A Grey to Blue Transition and the Evolution of Traditional SMR Process

Abstract:

Energy Transition is necessary to avoid climate changes, save the plant and, ultimately, save ourselves. It seems hydrogen is seen as one of the most important enablers to the energy transition as it burns without producing CO2 as by-product. However hydrogen is not a source of energy, it is a vector, it means that we don't have wells where to get it but, instead, we need to extract it from other species, kind of natural "hydrogen tanks": these hydrogen tanks are hydrocarbons and water.

The extraction of hydrogen from hydrocarbons is nothing new for the industry: refineries, petrochemical and chemical facilities use hydrogen since ages as it is vital for their processes, for this reason there are plenty of hydrogen units installed worldwide, most of them based on the Steam Reforming process. The hydrogen produced by these units is called "grey" and has the big challenge of producing 9 to 12 tons of CO2 as by-products for each ton of produced hydrogen, CO2 that is ultimately sent to the atmosphere. So, how can we make hydrogen at scale ensuring the reduction of CO2 emissions? One solution is the so called Blue hydrogen: hydrogen produced from fossil origin feedstock in which the CO2 is removed and sequestrated for re-use or storage

We hear a lot nowadays about blue hydrogen for new applications, however this is not enough to reduce the environmental impact: a big piece of the decarbonization plan pass through the need of reducing the carbon footprint of existing facilities, and existing hydrogen units are certainly one of the most important in the industry that need to be decarbonized.

This paper summarizes the job done by Wood to evaluate, technically and economically, the decarbonization options for a base case 100,000 Nm3/h grey hydrogen unit, using pre-combustion carbon capture, post-combustion carbon capture, integration of gas heated reactors as well as combination of all these solutions.

640443 Advanced Hydrogen Solutions in Decarbonizing Refineries

Abstract:

Hydrogen has been carrying its crucial role in petroleum refining for catering to the increasingly stringent requirements for cleaner and lighter product slate. Accordingly, it has drawn focal attention in terms of advanced hydrogen management as well as optimized on-purpose H2 generation for enhancing overall refinery profitability.

Over the years, the objective targets have progressed from H2 management towards CO2 management, more so since H2 generation itself caries quite a large C-footprint (of 9-12 ton per ton H2). Such efforts however led to only to a marginal reduction in CO2, while not governed by any statuary framework.

Based on the current and imminent global drive for tackling climate change, hydrogen is finding its newer evolving and expanded role as an enabler for the targeted energy transition. With the strong push towards 'near-zero' and then to 'net-zero' carbon targets, efforts are gearing up for the projected transforming of the conventional Grey H2 to Blue H2 (with added CO2 capture) with Green H2 (based on water electrolysis using renewable power) expected to catch up in the longer term.

However, the "hard-to-abate" industrial sectors comprising mainly of refining, petrochemicals, cement, DRI etc. which involve high temperature processing thus requiring fuel firing, are expected to follow the "blue pathways" in terms of fossil-continuity with CCS. With various continued technological developments in post-combustion capture, mainly to improve the energy and capital intensity as well as process reliability, there is no clear cut and cost-effective process which is sufficiently deployed commercially. Thus, considering the proposed decarbonization targets of > 80% for a Refinery with its multiple fuel-fired heaters, the underlying need for post-combustion CO2 capture makes it quite a challenging and perhaps risky venture on a larger scale.

To address such emerging needs, the advanced solutions for decarbonizing especially the hard-to abate set-up of a Refinery, wrest on the pathway of first switching the refinery-wide fuel gas network to low-carbon or carbon-free fuel such as hydrogen or its compounded form as ammonia and secondly to find ways to (cost) effectively decarbonize such H2 generation effectively while using conventional hydrocarbon feedstock.

In addressing the above trends, the paper will also describe the innovative and distinctive developments by ZoneFlow reactor technologies, firstly in terms of their structured catalytic ZF reactors for steam reforming which by replacing the existing pellets, allows a capacity increase of 15-20% combined with enhanced efficiency, thereby also lowering the CO2 per unit H2. And secondly as a bigger breakthrough, a rearranged conventional H2 plant flowsheet called " Z-H2" which allows >90% decarbonization only through deploying only the well-proven and cost-effective pre-combustion CO2 removal from high pressure clean syngas.

The Z-H2 concept is based on recycling and retaining all the carbon within the process while having the fuel for SMR firing as only H2 thereby fully eliminating the need for less desirable post-combustions capture. The larger H2 plant also provides the H2 fuel for rest of the refinery units replacing the refinery fuel gas thus achieving the targeted decarbonization along with centralized pre-combustion capture within the H2 plant, thus offering an effective way forward in addressing the imminent decarbonization targets.

AUTHOR INDEX